Clever Critters

HIBERNATION HOTSHOTS!

By Tracy Vonder Brink

TABLE OF CONTENTS

A Crabtree Seedlings Book

Crabtree Publishing

crabtreebooks.com

School-to-Home Support for Caregivers and Teachers

This book helps children grow by letting them practice reading. Here are a few guiding questions to help readers build comprehension skills. Possible answers appear here in red.

Before Reading:

- What do I know about this topic?
    - *I know that some animals hibernate during the winter.*
    - *I know that some animals hibernate underground.*
- What do I want to learn about this topic?
    - *I want to learn which animals hibernate.*
    - *I want to learn why hibernation is helpful.*

During Reading:

- What have I learned so far?
    - *I have learned that adaptations help living things survive.*
    - *I have learned that hibernation is a sort of deep sleep.*
- I wonder why...
    - *I wonder why a bear eats a lot in the summer and fall.*
    - *I wonder why the common poorwill is the only bird that hibernates.*

After Reading:

- What did I learn about this topic?
    - *I have learned that a toad digs a hole to hibernate.*
    - *I have learned that ladybugs hibernate together.*
- Read the book again and look for the glossary words.
    - *I see the word **survive** on page 3 and the word **soil** on page 6. The other glossary words are found on pages 22 and 23.*

HIBERNATION HOTSHOTS!

An **adaptation** is something that helps a living thing **survive**.

Some animals go into a sort of deep sleep to survive cold winters. We call this **hibernation**.

These animals are hibernation hotshots!

Many bears eat a lot during summer and fall.

They build up **fat** to be ready to hibernate.

They will not eat again until spring.

A toad digs into the ground when cold weather comes.

The **soil** keeps it warm like a blanket.

A queen bumblebee also digs into the ground.

Or she finds a hole in a tree.

She hibernates until warm weather comes.

Many birds fly south for the winter.

But the common poorwill hibernates.

It finds a spot on the ground or in some rocks.

The common poorwill is the only bird that hibernates.

A dormouse curls up in a nest.

It will not move at all while it hibernates.

A groundhog hibernates underground in a **burrow**.

It sleeps as long as the weather stays cold.

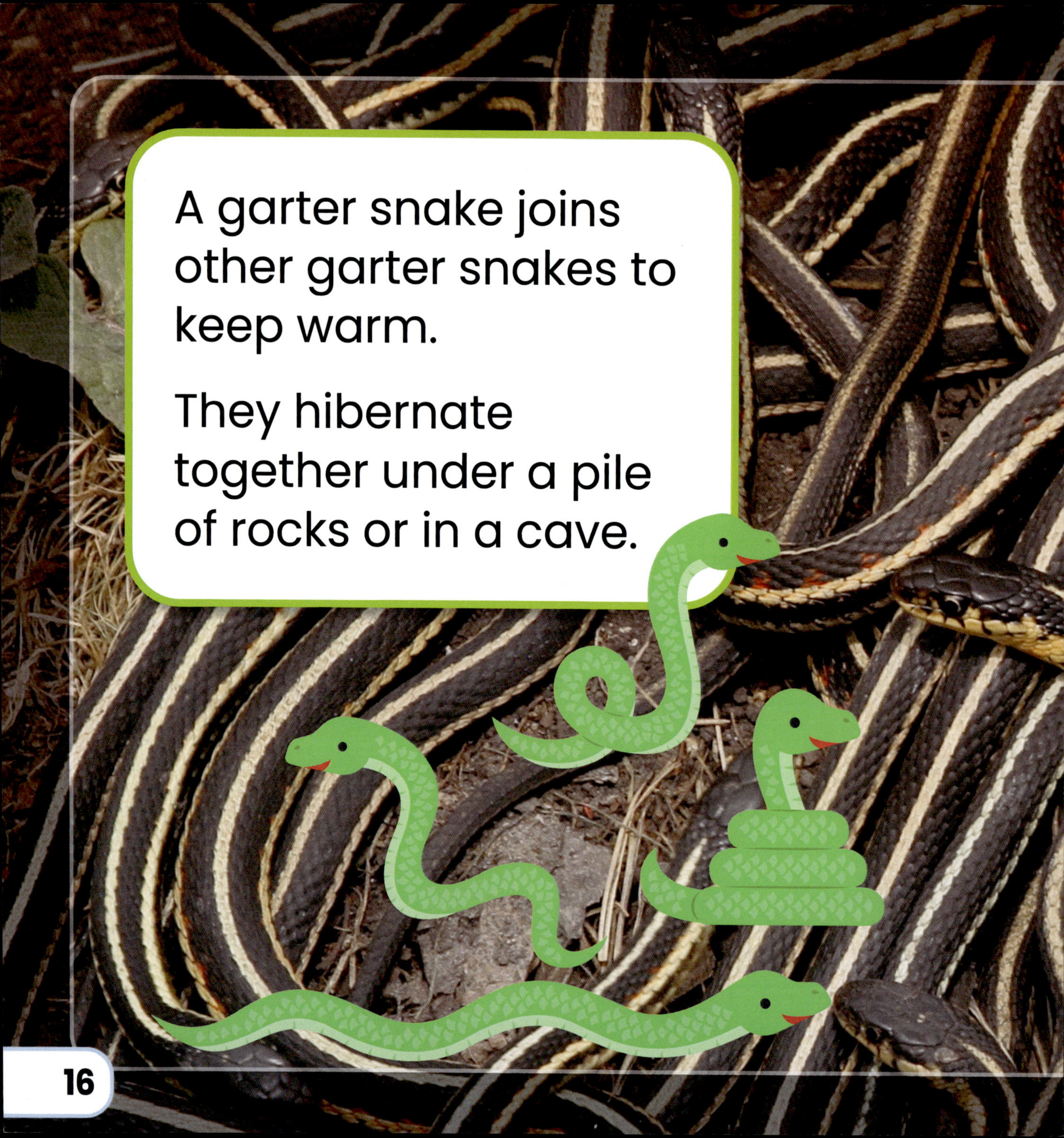

A garter snake joins other garter snakes to keep warm.

They hibernate together under a pile of rocks or in a cave.

Garter snakes sometimes hibernate in groups of 100 or more.

Ladybugs also hibernate together.

They gather under rocks or in an old log.

Thousands of ladybugs may hibernate together.

An Arctic ground squirrel digs a burrow.

It uses grass and fur to make a cozy nest for winter.

Hibernation is a useful adaptation for many living things!

Glossary

adaptation (ad-uhp-TEY-shuhn): A change in an animal or a plant that helps it live in a certain place

burrow (BUR-oh): A hole or tunnel that an animal digs in the ground to live in

fat (fat): A layer under the skin of an animal that keeps it warm

hibernation (hy-ber-NEY-shuhn): A deep sleep-like state that some animals go into to survive cold winters

soil (soil): The top layer of earth where plants grow

survive (ser-VAHYV): To continue to live

Index

About the Author

Tracy Vonder Brink loves to learn about science and nature. She has written more than 100 nonfiction books for kids. Tracy lives in Cincinnati, Ohio, with her husband, two daughters, and two rescue dogs.

Written by: Tracy Vonder Brink
Designed by: Margaret Salter
Proofreader: Melissa Boyce
Photographs: All images from Shutterstock

Crabtree Publishing

crabtreebooks.com 800-387-7650

Printed in the U.S.A./CP2025

Published in Canada
Crabtree Publishing
616 Welland Avenue
St. Catharines, Ontario
L2M 5V6

Published in the United States
Crabtree Publishing
347 Fifth Avenue
Suite 1402-145
New York, New York, 10016

Library and Archives Canada Cataloguing in Publication
Available at Library and Archives Canada

Library of Congress Cataloging-in-Publication Data
Available at the Library of Congress

Hardcover: 978-1-0398-8110-5
Paperback: 978-1-0398-8471-7
Ebook (pdf): 978-1-0398-8230-0
Epub: 978-1-0398-8350-5
Read-Along: 978-1-0398-8555-4
Audio: 978-1-0398-8513-4